ISBN 978-3-749-42055-1

Das Buch ist sorgfältig erarbeitet. Dennoch übernimmt der Autor, der Verlag und der Herausgeber in keinem Fall für die Richtigkeit von Angaben, Hinweisen und Ratschlägen sowie für eventuelle Druckfehler irgendeine Haftung

Autor : Dr. Hans-J. Dammschneider

Bibliographische Information:
Die Deutsche Nationalbibliothek verzeichnet diese Publikation in der Nationalbibliographie; detaillierte bibliographische Daten sind im Internet über http://www.dnb.d-nb.de abrufbar.

Alle Rechte, insbesondere die der Übersetzung in andere Sprachen, vorbehalten. Kein Teil dieses Buches einschliesslich der Abbildungen bzw. Grafiken darf ohne schriftliche Genehmigung des Verlages *und* des Autors in irgendeiner Form – durch Photokopie, Mikroverfilmung oder irgendein anderes Verfahren – reproduziert oder in eine von Maschinen, insbesondere von Datenverarbeitungsmaschinen, verwendbare Sprache übertragen oder übersetzt werden. Die Wiedergabe von Warenbezeichnungen, Handelsnamen oder sonstigen Kennzeichen in diesem Buch berechtigt nicht zu der Annahme, dass diese von jedermann frei benutzt werden dürfen. Vielmehr kann es sich auch dann um eingetragene Warenzeichen oder sonstige gesetzlich geschützte Kennzeichen handeln, wenn sie nicht eigens als solche markiert sind.

© : 2019 Inst.für Hydrographie, Geoökologie und Klimawissenschaften, Dr. Hans-J. Dammschneider

Herstellung
und Verlag : BoD – Books on Demand, Norderstedt

ISBN : 978-3-749-42055-1

Auflage : 2019-1

website : www.ifhgk.com

Abstract

Temperature changes of weather stations in Northern-Europe are evaluated for the period 1913-2013 and results are compared with the oceanic cycles of the Atlantic Multidecadal Oscillation (AMO) and the Pacific Decadal Oscillation (PDO) as well as the developments of the sea surface temperatures (SST). It is demonstrated that PDO and AMO as distribution patterns of potential oceanic thermal storage and heat release are closely correlated to the nearly in sync changing trends of air temperatures in Europe. In correlation to the warmer or colder conditions it can be observed in Europe that air temperatures are ascending or descending. There are spatial changes of temperatures in the oceanic oscillations of both the Pacific (PDO) and the Atlantic (AMO), their interplay and the dependend transport of latent allochthonous energy are a possible reason for the also ´periodic´ temperature trends in Europe ... whether for example in Bergen (N), Germany (D) or water temperatures in the North Sea (Helgoland, D).

Oceanic cycles and the variability of air and water temperatures in Northern-Europe

Hans-J. Dammschneider

		page
1	**Introduction**	8
2	**Oceanic cycles**	8
3	**Periodicities of air temperatures in Europe**	11
4	**Atmospheric circulation and transport of latent energy**	14
5	**Conclusion**	18
6	**Literature**	19
	Methodological note	25

1 Introduction

Climate is never constant, but has changed continuously since the beginning of the Earth's atmosphere. Most conspicuous are decadal climate fluctuations, including the so-called ocean cycles, which already existed in pre-industrial times and are still active today.

This contribution investigates the impact of the Pacific Decade Oscillation (PDO) and the Atlantic Multidecade Oscillation (AMO) on fluctuations in European air temperatures. The relative regularity of these multidecadal ocean cycles is largely independent of human influence and is primarily due to influences of natural origin from various sources. Positive and negative Phases alternate. The AMO has been reconstructed back to 1250 AD (KUHNERT and MULITZA, 2011), the PDO back to 993 AD (MACDONALD and CASE, 2005). Another prominent cycle is the North Atlantic Oscillation (NAO, e.g. WASSENBURG et al. 2016) which however is not topic of this contribution. The study period covers the last 100 years, i.e. commences in the early part of the early 20th century.

Multidecadal ocean cycles are superimposed on longer-term processes, requiring the analysis of at least 2 cycle periods before long-term climate trends can be confidently identified. The oceanic oscillations are most likely driven by large-scale and long-periodic redistributions of energy/heat, which takes place as part of the global oceanic circulation.

2 Oceanic cycles

Periodic changes or cycles of temperature distributions are best known from the Pacific (PDO/ENSO) and Atlantic (AMO). Interestingly, more attention has been paid to the patterns of these spatio-temporal oceanic oscillations than to those that also occur simultaneous and parallel in the atmospheric temperature processes of Europe for example.

This may be due to the fact that the "up and down" movements taking place in the european air temperatures are not as clearly recognisable at first glance as the trends that develop in the oceans. But, that is decisive, they also exist ... and if you record air temperatures in Europe (see e.g. fig. 2), they even have a frequency comparable to PDO and AMO. These cycles are apparently more than just random temperature changes, as we observe them in the usual distribution all over the world from year to year; they have trends.

Cycles are subject to complex processes and the oceanic periods are in this sense "patterns" of an energy distribution that shifts in intensity and spatial distribution over decades. These temperature intensity patterns are calculated as an index and change in space and time. The PDO has a mean period of 60 years. The cycle is currently in a downward trend, except for a very strong El Nino that influenced the PDO values in 2014-2016 (therefore the period under consideration ends (initially) in 2013, but does indeed include previous El Nino). Similarly, the AMO rose from 1920 to 1940, fell from 1940 to 1970 and then rose again strongly up to 1998 (see Fig. 1).

The concept of the PDO was developed by MANTUA (1997). The PDO has positive values when the interior of the North Pacific is abnormally cool and the eastern Pacific coast is warm. If the climate anomaly patterns are reversed, then the PDO has a negative value.

Already TISDALE (2009) noted that global temperatures rise particularly strongly when the Pacific Decadal Oscillation (PDO) is positive. In contrast, global climate typically stagnates or cools when the PDO is negative. Hence, the PDO appears to modulate the multidecadal temperature development in the world.

The PDO does not represent the SST, it is a dimensionless pattern of SST anomalies in the North Pacific. But it is nevertheless that PDO can be used as an indicator of temperature changes in the Pacific region as well as the actual SST´s (see Fig. 2 compared to Fig. 1 below). No doubt that the PDO was not originally calculated to explain atmospheric temperature fluctuations, it is just a pattern of SST variability and does not represent the SST itself. But one cannot ignore that the strength of PDO and the variability of the dimension of PDO together result in a measure which then correlates with the actual atmospheric temperatures (via the USA to Europe).

Fig.1 also shows that the SSTnh (30-60N) correlates well with the PDO. This means that it is not only the "patterns" of PDO that can be compared with the variability of European air temperature, it is also (naturally) the water temperatures (and their trends) themselves.

Unfortunately, the physics and internal dependencies on the formation of oceanic cycles are still not fully understood. Previous research showed that the sunspot cycle can influence the cyclicity of PDO and AMO (VAN LOON and MEEHL, 2014). ZHOU and others (2014) also recognize an influence of the sun on oceanic cycles and the NAO (both on a daily and annual scale).

Ultimately, there is still no certainty about the causes/drive of oceanic cycles. DIJKSTRA et al. (2006) formulated the thesis of a periodically disturbed thermohaline circulation (and even brought other models into play). More recently BELLOMO et al. (2016) describe a climate amplifier, which should show that the AMO to a certain extent is linked to changes in cloud cover with up to one third of AMO-associated temperature changes due to cloud effects. This means that there is a positive feedback between the total amount of clouds, the surface temperature (SST) and the atmospheric circulation, which increases the persistence and amplitude of the tropical branch of the AMO. Using numerical simulation, BELLOMO conclude, that ´cloud feedback´ can account for between 10% and a maximum of 31% of the observed SST anomalies associated with AMO over the tropics.

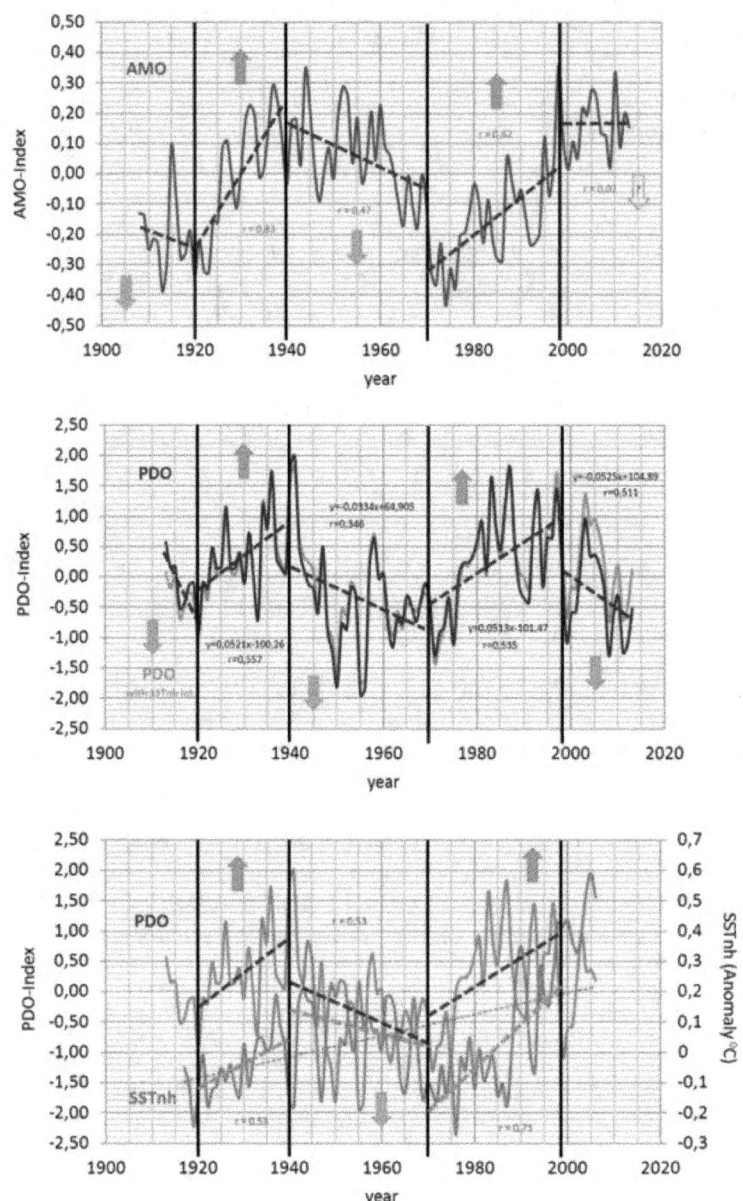

Figure 1: AMO (above), PDO (middle) and SSTnh (below), each with trend per period. All trend lines are linear regressions per time period. "PDO" means the original data (http://jisao.washington.edu/pdo/PDO.latest), for the "PDO with SSTnh int." the global warming signal was recalculated to the PDO from the SSTnh (means ´PDO **with** global warming´).

Numerical models continue to have problems forecasting oceanic cycles. This is due to the limited understanding which processes drive the variability of the cyclic processes of oceanic temperatures. Consequently, the cycles are numerically not yet fully describable.

ALEXANDER et al. (2008) have tried to solve the problem with the help of statistical methods, but with little success and a sample or condition description covering a maximum of 4 seasons (2 years, so without value forecast).

MEEHL et al. (2014) describe the problems that numerical models have with the representation/reproduction of ´Hiati´, which could be related to the Pacific oscillation.

KRAVTSOV at al. (2014) also come to the conclusion that numerical models are not yet able to depict the oceanic cyclicities.

KRAVTSOV et al. (2017) recently came to the conclusion that numerical models describe neither amplitude nor spatial distribution patterns of oceanic cycles with sufficient accuracy. The authors conclude that the models in their current form are hardly suitable for reproducing the real temperature curve.

3 Periodicities of air temperatures in Europe

Figure 2 shows the temperature development in Germany for the period since 1913, which is characterized by warming of about +0.9 degrees C. However, the temperature fluctuations are comparatively high from year to year at up to more than 2 degrees Celsius.

In climate research, the reasons for the medium- and short-term temperature modifications are mainly seen in the NAO (see RUPRECHT, 2008, or REINTGES, LATIF et al., 2016). The latter could represent a very close coupling of 8-year fluctuations in the temperature behaviour of the North Atlantic and the air temperatures. According to those authors, the primary influence of the NAO on Northern Europe is so strong that the period is even reflected in Hamburg's average winter temperatures.

A direct connection between the development of the oceanic cycles/the SSTnh and the changes in atmospheric temperatures could be expected, but this has rarely been discussed so far. However, even a simple graphical comparison of the course of PDO/SSTnh and that of Germany's air temperatures shows in Fig. 2 and Fig. 3, respectively, that for each period of the above PDO/SSTnh cycles, the trends in atmospheric temperatures also follow the same direction. Due to the fact that the PDO is only an index, it is of course not to be expected that the degree of temperature changes corresponds to that of the PDO. Nevertheless, the fundamental tendencies for each period are remarkable.

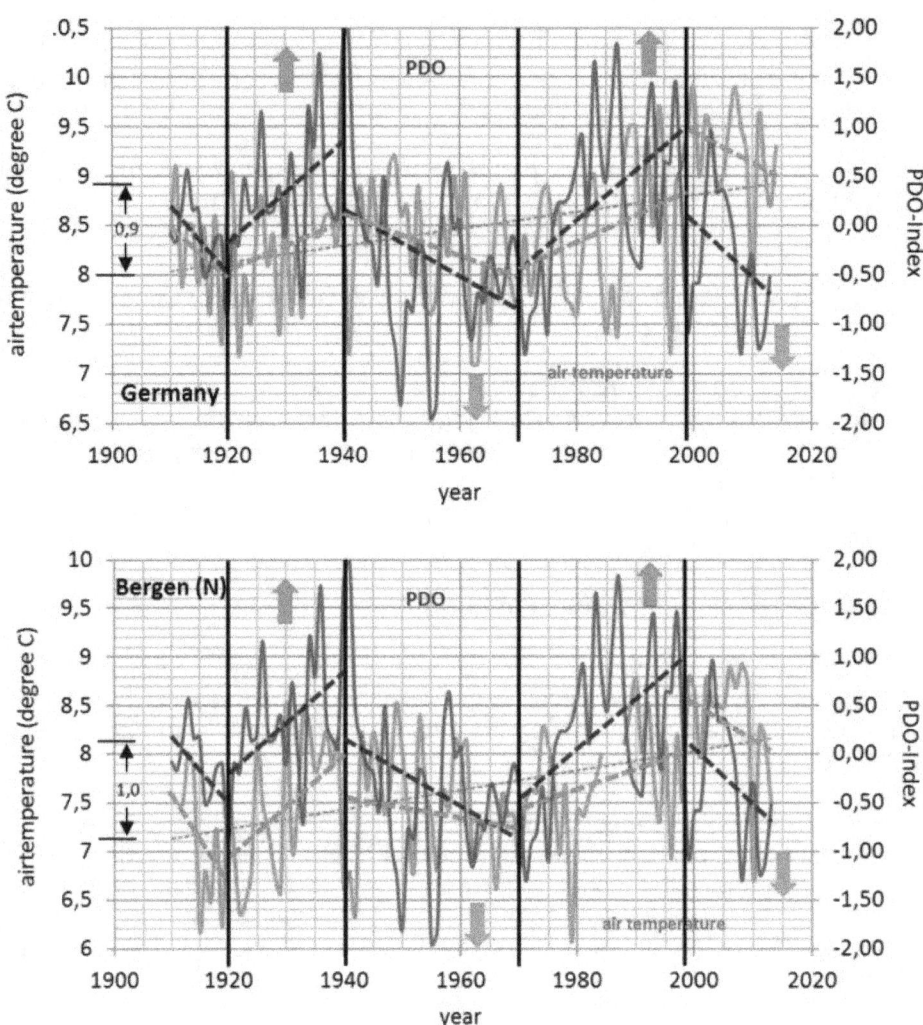

Figure 2: PDO and air temperature with trends per time unit, Germany (top) and Bergen (bottom). Data of the PDO see Fig. 1, Air temperatures Germany to DWD (https://www.dwd.de/DE/leistungen/zeitreihenundtrends/zeitreihenundtrends.html), air temperatures Bergen to GISS (https://data.giss.nasa.gov/cgi-bin/gistemp/stdata_show.cgi?id=646067000000&dt =1&ds=5). Note: The evaluation of the stations ABERDEEN (GB), NANTES (F), GENEVA (CH), COPENHAGEN (DK), POTSDAM (D), GENUA (I) and others were carried out by the author according to the same criteria and each station shows in principle/in trends the same results.

As already emphasized, the index of the PDO is not a temperature curve. Drawing direct conclusions from the positive and negative periods on the corresponding temperature curves, e.g. in Europe, is initially only a simplification in the form of a graphic observation.

But of course there can also be no doubt that with atmospheric and global circulation there is a global redistribution of energy ... a PDO value or SSTnh values (Fig. 1) above average for years must have consequences for the spatial (shifting with the west wind drift) distribution of heat from the respective area of the Pacific. It should therefore not be regarded as a coincidence that (especially depending on the spatial distribution of SST´s = PDO/AMO) there would be an impact on more distant areas of the northern hemisphere as far as Europe.

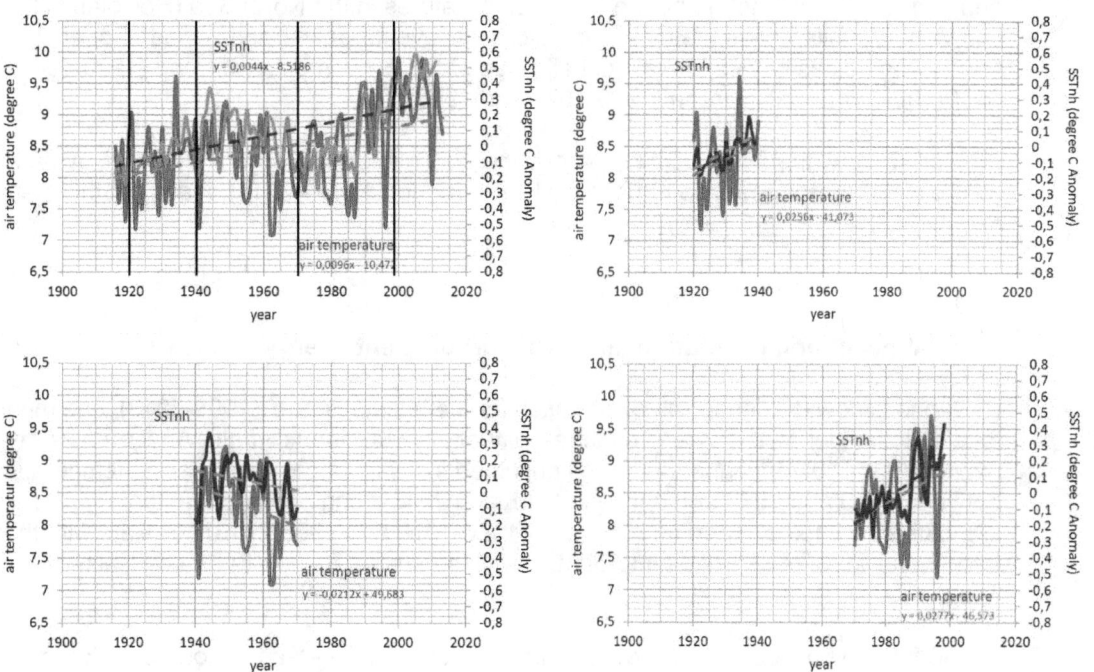

Figure 3: SSTnh and air temperatures in Germany with time period representations 1920-1940, 1940-1970 and 1970-1998 (data see Fig. 1).

The graphs show that there are strong similarities between the positive and negative phases of PDO/SSTnh and the trends of air temperatures in Germany (overall) and Bergen (N, example for single station in Northern Europe) that can be observed at the same time. This also applies to the water temperatures in the North Sea (Fig. 4, Helgoland): If the PDO/SSTnh is in a positive rising phase, then the temperatures in the North Sea also tend "upwards", but if the PDO/SSTnh falls into a negative falling phase, then Helgoland's water temperatures are also on a "downwards" trend.

The general observation that the temperature increase in Europe is by no means linear and uniform is nothing new. However, it has not yet been explained in more detail that the connection between PDO and the trends in air temperatures is much closer than one might think when looking at the entire period at first glance. Only the inclusion of the oceanic cycles gives the same picture for the air temperatures in Europe that we already know from PDO. According to BRÖNNIMANN (2015), the statement that the temperature development is "staircase-shaped" is not tenable according to the present representations in Figs. 2 to 4.

Comparison of the trend values from PDO, AMO, SSTnh, air temperatures (Bergen/N and Germany as a whole) and the water temperatures in the North Sea (Helgoland/D) shows how similar these stations and indices behave for each trend period: For each period, regardless of location, almost all values tend to be in positive trend (1920-1940 and 1970-1998) or negative trend (1940-1970 and 1998-2013).

Surprisingly, the same applies even to the water levels in Bergen (N): In the same trend as the air temperatures (see Fig. 2), the RMSL (period 1940 - 2013) of at least this station rise and fall (see Fig. 4 below).

4 Atmospheric circulation and transport of latent energy

If one is now looking for an explanation for the latest ´Hiatus´ (1998-2013), one must take a closer look at the overall changeability of the European climate. Analyzing its rhythm over the last 100 years and the trends observed in it with occasionally falling temperatures between 1940 and 1970 and between 1998 and 2013 it can be seen that the standstills/declines of warming in Europe (1940-1970 and 1998-2013) correlate with the trend behavior of the Pacific and Atlantic oscillation (see Fig. 2 to 5) ... a causality remains open.

So far, the publications of recent years have only interpreted the course of European temperature development as a rising "staircase", i.e. periods of temporarily lower warming are followed by periods of relatively stronger temperature increase. This rating can be found at BRÖNNIMANN (2015) or YU&XIE (2016). However, all basic data presented in Figs. 2 to 4 do not show ´stairs´, but rather periodicities in which (at least in Europe) rising temperatures alternate temporarily with falling temperatures ... one should certainly regard this as climate fluctuations within climate change.

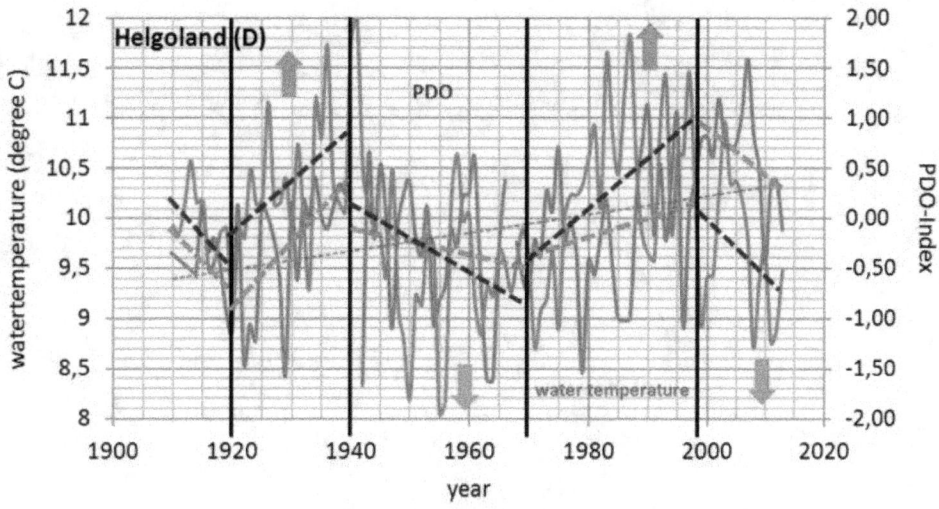

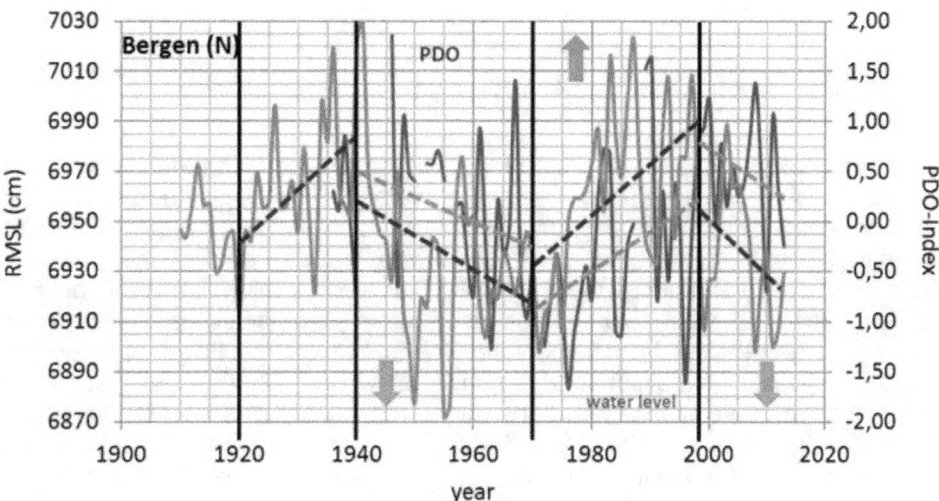

Figure 4: History curve of PDO and water temperatures HELGOLAND (D, graphic above) and course of PDO and water level BERGEN (graphic below), all with trends per time unit. The time series after 1998 ends in 2013 and this applies to all stations presented in this publication. The reason for this is that an unusually strong El Nino occurred in 2014-2016, which led to unusual values in all temperature recordings worldwide. The period under consideration here therefore deliberately ends in 2013. At this point, special thanks are due to Mr. Frohse of the BSH in Hamburg, who provided the North Sea water temperature values, which are also recorded by the AWI (Prof. Dr. Wiltshire) or which the author evaluated (1920-2000 = table, 2000-2013 from http://www.bsh.de/de/Meeresdaten/ Beobachtungen/ Meerestemperaturen_und_Waermeinhalte/helgoland.jsp).

periode	PDO with SSTnh int.	SSTnh 30-60	temperatures air Bergen (1)	Germany (2) degrees C	water Helgoland (3)	
1920 – 1940	+ 1,1	+ 0,2 °C	+ 0,9 °C	+ 0,5 °C	+ 1,2 °C	trends per periode
1940 – 1970	- 1,1	- 0,2 °C	- 0,6 °C	- 0,6 °C	- 0,4 °C	
1970 – 1998	+ 1,8	+ 0,4 °C	+ 0,6 °C	+ 0,8 °C	+ 0,6 °C	
1998 – 2013	- 0,6	+ 0,3 °C -0,05 °C at HadSST2-nh	- 0,5 °C	- 0,5 °C	- 0,8 °C	
overall trend	+ 0,3	+ 0,5 °C	+ 1,0 °C	+ 0,9 °C	+ 1,0 °C	

data according to (1) GISS ; (2) DWD ; (3) BSH/AWI ; PDO and SSTnh see figures 1 and 2

Figure 5: PDO, SSTnh and temperatures 1920-2013, trends per time unit in °C / index

These trends are far less due to European (autochthonous) atmospheric conditions, but rather allochthonous processes from oceanic (temperature) cycles play a role. Their storage/buffering capacity, internal dynamics and ultimately the long-range redistribution of heat emanating from them must lead to a more or less variable transition (heat-flux) of energy between ocean and atmosphere. The IPCC stated in 2007: "During the last 50 years, net heat fluxes from the ocean to the atmosphere demonstrate locally decreasing values (up to 1 W m-2 yr-1) over the southern flank of the Gulf Stream and positive trends (up to 0.5 W m-2 yr-1) in the Atlantic central subpolar regions (Gulev et al., 2006). At the global scale, the accuracy of the flux ob-servations is insufficient to permit a direct assessment of changes in heat flux" (Bindoff, N.L. et al. 2007, IPCC, Section 5.2.4). The long-term storage capacity of oceanic water masses speaks for a medium-term influence of the water temperatures on the atmosphere/moving atmosphere above them.

According to Kappas (2009), "as is well known, differences between precipitation and evaporation levels on Earth must be compensated for by meridional water vapour transports. The water vapour is transported from the areas with excess evaporation both to the pole and to the ITC. This compensatory movement is associated with a considerable transport of latent energy".

Kappas (2009) discusses whether too little attention is paid to this transport of latent energy in the system of atmospheric circulation and its effects on climatic fluctuations. This means that if the importance of the heat flux "sea - air" is underestimated or long-term climate change is indeed a phenomenon that is also based on the (long-term) energy transition from the atmosphere to the ocean, then the climate fluctuations (with PDO/AMO) are, however, a process that takes up changes from oceanic conditions (cycles) in the medium term and redistributes them by atmospheric circulation, having a distant effect all the way to Europe.

GULEV, LATIF at a. (2013) show in their work that there is a striking heat flux in the North Atlantic. Measurements of OHC and ARGO in the range between 35/50 degrees N were evaluated, which prove that the ocean currents influence the surface temperature of the Atlantic and thus above all the heat exchange with the atmosphere, which in turn causes climate fluctuations on the adjacent continents via the transport of latent energy.

KLÖVER, LATIF at a. (2014) reconstruct the heat flux between 1900 and 2010 from known data of the NAO in order to work out the decadal fluctuations of the North Atlantic SST. The data were used to attempt numerical predictions of the expected AMO progression. The result was what exactly actually had been observed since 1998 could happen, namely a negative trend.

According to HAIJUN YANG et al. (2015), who investigated the meridional heat transport (MHT) in the climate system with a coupled climate model (CESM1.0), the heat transport (OHT) is carried by both the wind-driven circulation and the thermohaline circulation. It dominates the total OHT in the mid-high latitudes. The work of the authors confirms that the meridional transport of atmospheric heat results primarily from the transport of latent energy (LE). Especially in the outer tropics, the latent heat transport tends towards the poles in the course of the west wind drift. The work also shows that it is even possible to quantify the contribution of individual marine areas to latent heat transport.

According to KIM, WEBSTER and CURRY (2012), however, the accuracy of numerical models is still limited in a wide range. Especially the PDO has a relatively poor prediction accuracy in most models. This is, therefore, a problem when exactly the spatial changes associated with the PDO of 'heat content' are suspected of also being involved in the cyclical temperature fluctuations of Europe/North America via the quantitatively-spatially varying transport of latent energy. However, if the processes of PDO cannot really be predicted with sufficient accuracy, a forecast of the transport of latent energy from the Pacific is still not available.

A comparison between the trends of PDO and those of the European-North American air temperatures suggests that the two are interrelated. If the values of PDO/AMO/SSTnh rise, the air temperatures in Europe also rise. If the values of the oceanic cycles/SSTnh fall, then the air temperatures in Europe also decrease. As already mentioned, the overall rise in global temperature is seen by some authors as "stairs" (e.g. BRÖNNNIMANN 2015 and YU&XIE 2016). It should, however, not be forgotten that this description is less suitable for European air temperatures, because no 'staircase' constantly climbs upwards, but a temperature "wave" runs upwards along a flat ascent.

In this context, it is interesting that the fluctuations in air temperatures seem to be more closely related to the changes in oceanic temperature distribution patterns (PDO/AMO) than to SST´s itself ... not only oceanic temperatures seem to play a role in the periodicity of European air temperatures, but also the spatial changes/distribution of periodic higher/lower temperatures in the Pacific and Atlantic, expressed as PDO and AMO.

In fact, MEEHL et a. (2016) assume that an important mode in global air temperature variability is the internal variability of PDO, which for example contributed to a reduced GMST trend in the early 2000s. Or to put it another way: The oceanic cycles influence the fluctuations in air temperatures worldwide. Long-term, externally forced trends of global mean surface temperatures (GMSTs) are embedded in the background noise of internally generated multi-decadal variability and these are by no means continuously increasing, but also have ´breaks´ with a temporarily decreasing trend.

YU & XIE (2016) identify four events in which tropical Pacific decadal cooling has significantly slowed the global warming trend. According to this, the tropical Pacific is the "key pacemaker" of the warming trend they have described as ´staircase´. SHUAI-LEI et a. (2017) also show that the globally averaged surface temperature has significant multidecadal fluctuations, which are characterized by two periods with slight decrease and two periods with stronger temperature increase. The results of their numerical model runs indicate that the SST of the Pacific play a major role in changes in global air temperatures. Or more precisely, multi-decadal global warming rate changes are primarily attributed to changes in ocean surface temperatures.

After all, 72% of the earth's surface is occupied by the oceans and in this area the vertical transport of latent heat plays the greater role, i.e. it participates with approx. 80 to 85% of the global heat balance (annual sum of the indirect lateral heat transport according to data from SELLERS 1965 and GRAEDEL&CRUTZEN 1994). According to this, almost 88% of the radiation balance at sea is used for the transport of latent heat and only 12% for the direct heating of the air.

5 Conclusion

PDO and AMO as distribution patterns of potential oceanic thermal storage and heat release show significant oscillations which are also represented in the air temperatures in Europe. The ups and downs of the climate (e.g. Bergen/N) correlates closely with the course of the oceanic cyclicities in the Pacific (PDO) and the Atlantic (AMO). This also applies to the water temperatures in the North Sea (Helgoland/D). Taking into account the atmospheric circulation and the transport of latent energy, there is a ´long-distance impact´ of the oceanic cycles that includes Europe, as reflected in the periodicity of European air temperatures.

Focus of this study is on observations ... the relationship between the changes or cyclicities of the oceanic PDO/AMO and the oscillations of the European temperatures is, to say it simple, based on the assumption that the basic rules of physics and energy transport apply. Consideration of the effects of atmospheric circulation on the transport of latent energy is therefore a logical derivation from generally known and recognised climatological facts. However, it is certain that the overall periodicity of air temperatures cannot directly result from changes in the near-field atmosphere; it is imperative to assume a remote effect.

A medium- to long-term climate change is modulated by these natural oscillations. The question is how the occurrence and regularity of the cycles of European air temperatures are to be interpreted. One cannot avoid having to look for the causes also in the internal thermal-oceanic upheavals of the Pacific or Atlantic. There are changing intensities of latent energy which are available from the oceans via ´heat flux´. Because of the changing patterns in the areal oceanic energy distribution the indices of PDO and AMO seems to be visible in the fluctuations of European air temperatures.

6 Literature

Alexander, M.A. et.a. (2008): Forecasting Pacific SSTs: Linear Inverse Model Predictions of the PDO. In: Journal of Climate, Vol. 21, 2, 2008

Balmaseda, M.A., Trenberth, K.E. and Källen, E. (2013): Distinctive climate signals in reanalysis of global ocean heat content. In: Geophysical Research letters, Vol. 40, 2013

Barcikowska, M.J. et a. (2016): Observed and simulated fingerprints of multidecadal climate variability, and their contributions to periods of global SST stagnation. In: AMS, 2016

Becker, G.A., Frohse, A. & Damm, P. (1997): The northwest european shelf temperature and salinity variability. In: Deutsche Hydrographische Zeitschrift, 49, 1997

Bellomo, K. et a. (2016): New observational evidence for a positive cloud feedback that amplifies the Atlantic Multidecadal Oscillation. In: Geophysical Research Letters, Vol. 43, 2016

Bindoff, N.L. et a. (2007): Observations: Oceanic Climate Change and Sea Level. In: Climate Change 2007: The Physical Science Basis. Contribution of Working Group I to the Fourth Assessment Report of the Intergovernmental Panel on Climate Change, IPCC 2007

Biondi, F. et a. (2001): North Pacific Decadal Climate Variability since 1661. In: Journal of climate, Vol. 14, 1, 2001

Brönnimann, S. (2015): Climate science - Pacemakers of warming. In: Nature Geoscience 8, 2015

Bubenzer, O. and Radtke, U. (2007): Natürliche Klimaänderungen im Laufe der Erdgeschichte. In: Der Klimawandel – Einblicke, Rückblicke und Ausblicke (Klimawandel). Humboldt-Universität, Berlin 2007

Cheng, W. et a. (2013): Atlantic meridional overturning circulation (AMOC) in CMIP5 models: RCP and historical simulations. In: J. Climate, 26, 7187– 7197.

Chikamoto, Y. et a. (2016): Potential tropical Atlantic impacts on Pacific decadal climate trends. In: Geophysical Research Letters, 43, 2016

Chylek, P., Dubey, M.K., Lesins, G. (2014): Imprint of the Atlantic multi-decadal oscillation and Pacific decadal oscillation on southwestern US climate: past, present, and future. In: Climate Dynamics, 43, 2014

Compo, G.P. and Sardeshmukh, P.D. (2009): Oceanic influences on recent continental warming. Climate Dynamics, 32, 333-342

D´Aleo and Easterbrook, D. (2011): Relationship of Multidecadal Global Temperatures to Multidecadal Oceanic Oscillations. In: Evidence Based Climate Change Series, 2011, p.161-184

Delworth, T.L., Mann, M.E. (2000): Observed and simulated multi-decadal variability in the Northern Hemisphere. In: Clim. Dyn. 16,2000

Dijkstra, H.A., te Raa, L., Schmeits, M. et al. (2006): On the physics of the Atlantic Multidecadal Oscillation. In: Ocean Dynamics 56, 2006

England, M.H. et a. (2014): Recent intensification of wind-driven circulation in the Pacific and the ongoing warming hiatus. In: nature climate change, published online 9 February 2014

Galbraith, E.D. et a. (2007): Carbon dioxide release from the North Pacific abyss during the last deglaciation. In: Nature, 449, 2007

Gouretzki, V. et a. (2012): Consistent near-surface ocean warming since 1900 in two largely independent observing networks. In: Geophysikal Research Letters 39, 2012

Graedel, T.E & Crutzen, P.J. (1994): Chemie der Atmosphäre. Heidelberg 1994

Gulev, S.K., Latif, M. et a. (2013): North Atlantic Ocean Control on Surface Heat Flux at Multidecadal Timescales. In: Nature, 499, 464-467, doi: 10.1038/nature12268

Han, Z. u.a. (2016): Simulation by CMIP5 models of the atlantic multidecadal oscillation and its climate impacts. In: Advances in Atmospheric Sciences, v. 33, no. 12, p. 1329-1342.

Haijun Yang, Qing Li, Kun Wang, Yu Sun and Daoxun Sun (2015): Decomposing the meridional heat transport in the climate system. In: Climate Dynamics, Vol 44, Issue 9, 2015

IPCC (2014): Fifth Assessment Report, AR5, Genf 2014

Kappas, M. (2009): Klimatologie. Klimaforschung im 21.Jahrhundert. Springer Verlag, Heidelberg 2009

Kim, H.M, Webster, P.J. & Curry, J.A. (2012): Evaluation of short-term climate change prediction in multi-model CMIP5 decadal hindcasts. In: Geophysical Research Letters, Volume 39, Issue 10, 2012

Klöver, M., Latif, M. et a. (2014): Atlantic meridional overturning circulation and the prediction of North Atlantic sea surface temperature. In: Earth and Planetary Science Letters, 406, 2014

Knudsen T.R. et a. (2016): Prospects for a prolonged slowdown in global warming in the early 21st century. In: Nature Communication 7, 2016

Kravtsov, S. et a. (2014): Two contrasting views of multidecadal climate variability in the twentieth century. In: Geophysical Research Letters, 41, 2014

Kravtsov, S. et a. (2017): Pronounced differences between observed and CMIP5-simulated multidecadal climate variability in the twentieth century. In: Geophysical Research Letters, 44, 2017

Kunert, H. and Mulitza, S. (2011): Multidecadal variability and late medieval cooling of near-coastal sea surface temperatures in the eastern tropical North Atlantic. In: Paleoceanography, 26, 2011

Latif, M., Roeckner, E. et a. (2004): Reconstructing, Monitoring, and Predicting Multidecadal-Scale Changes in the North Atlantic Thermohaline Circulation with Sea Surface Temperature. In: Journal climate, 17, 2004

Latif, M., Boening, C.W. et a. (2006): Is theThermohaline Circulation Changing? In: Journal of Climate, 19, 2006

Levitus, S. et a. (2005): Warming of the world ocean, 1955– 2003. In: Geophysical Research Letters, Vol. 29, 2005

Levitus, S. et a. (2012): World Ocean heat content and thermosteric sea level change (0-2000m), 1955-2010. Geophys.Res.Letters 39, 2012

Li, Andy K. et a. (2016): The changing influences of the AMO and PDO on the decadal variation of the Santa Ana winds. In: Environ. Res. Lett. 11, 2016

Lyman, J.M. & Johnson, G.C. (2014): Oceanography: Where's the heat? In: Nature Climate Change 4, 956– 957, 2014

Mantua, N.J. et a. (1997): A Pacific Interdecadal Climate Oscillation with Impacts on Salmon Production. In: Bulletin of the American Meteorological Society 78(6), 1997

Mantua N.J., Steven R. Hare (2002): The Pacific Decadal Oscillation. In: Journal of Oceanography. Volume 58, Nr. 1, 2002, S. 35– 44

MacDonald, G.M. and R.A. Case (2005): Variations in the Pacific Decadal Oscillation over the past millennium. In: Geophys. Res. Lett., 32, 2005

Mauritzen, C. et a. (2012): Importance of density-compensated temperature change for deep North Atlantic Ocean heat uptake. In: Nature Geoscience, 10, 2012

McCarthy, G.D. et a. (2015): Ocean impact on decadal Atlantic climate variability revealed by sea-level observations. In: Nature 521, 508– 510, May 2015

Meehl, G.A. et a. (2013): Externally Forced and Internally Generated Decadal Climate Variability Associated with the Interdecadal Pacific Oscillation. In: Journal of climate, Vol. 26, 2013

Meehl, G.A. et a. (2014): Climate model simulations of the observed early-2000s hiatus of global warming. In: Nature Climate Change, 4, 2014

Meehl, G.A. et a. (2016): Contribution of the Interdecadal Pacific Oscillation to twentieth-century global surface temperature trends. In: Nature Climate Change 6, 2016

Murphy, Lisa N., Bellomo, K., Cane, M. and Clement, A. (2017): The role of historical forcings in simulating the observed Atlantic multidecadal oscillation. In: Geophysical research letters, 44, 2017

Orgeville, Marc and Peltier, W.R. (2007): On the Pacific Decadal Oscillation and the Atlantic Multidecadal Oscillation: Might they be related? In: Geophysical Research Letters, Vol. 34, 2007

Petit, J.R. et a. (1999): Climate and atmospheric history of the past 420.000 years from the Vostok Ice core, Antarctica. In: Nature, 399, 1999

Reintges, A., M. Latif, W. Park (2016): Sub-Decadal North Atlantic Oscillation variability in observations and the Kiel Climate Model. In: Climate Dynamics, http://dx.doi.org/10.1007/s00382-016-3279-0

Roemmich, D. (2012): New Comparison of Ocean Temperatures Reveals Rise over the Last Century. In: Scripps Institution of Oceanography, San Diego 2012

Scafetta, N. (2013): Multi-scale dynamical analysis (MSDA) of sea level records versus PDO, AMO, and NAO indexes. In: Climate Dynamics, 43, 2013

Semenov, V., Latif, M., Dommenget, D. et a. (2010): The impact of North Atlantic– Arctic multi-decadal variability on Northern Hemisphere surface air temperature. In: J. Clim 23, 2010

Shen, C. et a. (2006): A Pacific Decadal Oscillation record since 1470 AD reconstructed from proxy data of summer rainfall over eastern China. In: Geophysical research letters, Vol. 33, 2006

Shuai-Lai, Yao et a. (2017): Distinct global warming rates tied to multiple ocean surface temperature changes. In: Nature Climate Change, 7, 2017

Speth, P. (1974): Horizontale Flüsse von sensibler und latenter Energie und von Impuls für die Atmosphäre der Nordhalbkugel. In: Met.Rundschau, 27, 1974

Tisdale (2009), pers. comm. , https://bobtisdale.wordpress.com/2009/04/27/misunderstandings-about-the-pdo-%E2%80%93-revised/.

Tootle, G.A., Piechota, Th.C. (2006): Relationships between Pacific and Atlantic ocean sea surface temperatures and U.S. streamflow variability. In: Water Resources Research, 42, 2006

van Loon, H. and Meehl, G.A. (2014): Interactions between externally forced climate signals from sunspot peaks and the internally generated Pacific Decadal and North Atlantic Oscillations. In: Geophysical research letters, 41, 2014

Wassenburg, J.A. et a. (2016): Reorganization of the North Atlantic Oscillation during early Holocene deglaciation. In: Nature Geoscience 9, 602-605, 2016

Willis, J.K. et a. (2004): Interannual variability in upper ocean heat content, temperature, and thermosteric expansion on global scales. In: Journal of Geophys.Research, Vol. 109, 2004
Wiltshire & Manly (2004): Helgol Mar Res 58, 2004

Wu, S., Liu, Z. et a. (2011): On the observed relationship between the Pacific Decadal Oscillation and the Atlantic Multi-decadal Oscillation. In: Journal Oceanography, 67, 2011

Wunsch, C. and Heimbach, P. (2014): Bidecadal Thermal Changes in the Abyssal Ocean. In: Journal of Physical Oceanography, 44, 8, 2014

Wyatt, M.G., Kravtsov, S. and Tsonis, A.A. (2012): Atlantic Multidecadal Oscillation and Northern Hemisphere's climate variability. In: Climate Dynamics, 38, 2012

Yu, K. & Xie, S.-P. (2016): The tropical Pacific as a key pacemaker of the variable rates of global warming. In: Nature Geoscience, 9, 2016

Zhang, R. and Delworth, T. (2006): Impact of Atlantic multidecadal oscillations on India/Sahel rainfall and Atlantic hurricanes. In: Geophysical research letters, Vol. 33, 2006

Zhou, L., Tinsley, B. and Huang, J. (2014): Effects on winter circulation of short and long term solar wind changes. In: Advances in space research, 54, 2014

Methodological note

The author deliberately does not make any mathematical-statistical correlations between the oceanic cycles and the trends in European air temperatures. Finally, a purely graphical comparison is made between the course of the combined PDO/AMO (by time period) and that of the air temperatures of continental stations in Europe and North America.

The trends are represented in their linear progression, the determination of the time intervals was carried out according to the relative maxima/minima of the oceanic cycles. The axis division of the graphs is in degrees C, the indices of the oceanic cycles are plotted in the same gradation as those of the air temperatures.

The comparative work of D´Aleo&Easterbrook (2011) was reproached because the AMO/PDO/air temperature correlations carried out by these authors were possibly inadmissible due to "smoothing" procedures. The author would therefore like to stress once again that the graphs listed in this text do something different, namely nothing more than to compare the respective linear trends of AMO/PDO with those of the air temperatures of selected stations.

Thus, as already emphasized, there is no mathematical correlation, but a purely graphical comparison ... the "narrowness" of a relationship is deliberately left to the eye of the beholder.

This is not a cause-and-effect comparison, but exclusively shows that there is a relationship between the oceanic cycles and the air temperatures of Europe.

For the selection of the observation periods etc. see also www.ifhgk.org (series of publications and there "Explanations to Volume 2").

www.ingramcontent.com/pod-product-compliance
Lightning Source LLC
Chambersburg PA
CBHW062209220526
45470CB00009B/2982